Unearthing the Secrets of Geological Engineering A Student's Perspective

Ricardo kane

Copyright © [2023]

Title: Unearthing the Secrets of Geological Engineering A Student's Perspective
Author's: Ricardo kane

This book was printed and published by [Publisher's: **Ricardo kane**] in [2023]

ISBN:

TABLE OF CONTENT

Chapter 11: Case Studies in Geological Engineering   67

Geotechnical Challenges in Construction Projects

Rehabilitation of Damaged Structures

Innovations in Geological Engineering

Chapter 12: Future Trends in Geological Engineering   73

Emerging Technologies in the Field

Sustainable Practices in Geological Engineering

Opportunities for Research and Development

Chapter 13: Conclusion and Final Thoughts   79

Reflecting on the Journey as a Geological Engineering Student

Advice for Future Geological Engineering Students

Continuing Education and Professional Growth in Geological Engineering

Chapter 1: Introduction to Geological Engineering

What is Geological Engineering?

Geological engineering is an exciting and multidisciplinary field that combines elements of geology, engineering, and environmental sciences. It is a branch of engineering that focuses on understanding and manipulating the Earth's resources and geological processes to design and build safe and sustainable infrastructure. In this subchapter, we will delve into the fascinating world of geological engineering, exploring its principles, applications, and the key skills required to succeed in this field.

At its core, geological engineering encompasses the study of the Earth's materials, structures, and processes. This knowledge is crucial for identifying potential hazards, such as landslides, earthquakes, and volcanic eruptions, and developing strategies to mitigate their impact on human activities. Geological engineers play a vital role in designing and constructing infrastructure projects, such as tunnels, dams, and foundations, which must withstand the dynamic forces of the Earth.

One of the key aspects of geological engineering is the understanding of geological formations and their behavior under different conditions. By analyzing the properties of rocks, soil, and water, geological engineers can assess the stability and suitability of a site for construction. They employ sophisticated techniques, including remote sensing, geophysical surveys, and laboratory testing, to gather data and make informed decisions.

Geological engineers also contribute to environmental conservation and sustainable development. They work on projects related to groundwater management, contaminated site remediation, and natural resource exploration. By implementing innovative techniques and technologies, they aim to minimize the environmental impact of human activities and ensure the responsible use of Earth's resources.

To pursue a career in geological engineering, students need to develop a strong foundation in mathematics, physics, and geology. They must also acquire skills in computer modeling, data analysis, and project management. Additionally, fieldwork is an integral part of geological engineering education, allowing students to gain hands-on experience in collecting and interpreting geological data.

In conclusion, geological engineering is a captivating field that offers a unique blend of scientific exploration and engineering solutions. As students interested in this discipline, you have the opportunity to contribute to society by ensuring the safe and sustainable development of our infrastructure and natural resources. By understanding the principles and applications of geological engineering, you will be well-equipped to embark on a rewarding and impactful career in this exciting field.

The Importance of Geological Engineering

As students exploring the field of geological engineering, it is crucial to understand the significance of this discipline and its impact on society. Geological engineering plays a vital role in various aspects of our lives, ranging from infrastructure development to environmental protection. In this subchapter, we will delve into the importance of geological engineering and why it is a promising career path for students interested in the intersection of geology and engineering.

One of the primary contributions of geological engineering lies in its ability to ensure the stability and safety of infrastructure projects. Whether it is the construction of buildings, bridges, or tunnels, geological engineers play a crucial role in assessing the geological conditions of the site. By conducting thorough investigations, they can identify potential hazards such as landslides, sinkholes, or earthquakes that may pose risks to the structure's integrity. Through their expertise, geological engineers develop effective strategies to mitigate these risks, ensuring the safety of the infrastructure and the people who utilize it.

Geological engineering also plays a pivotal role in environmental conservation and the sustainable use of natural resources. With the growing concerns over climate change and the depletion of finite resources, it is imperative to employ sustainable practices in various industries. Geological engineers are instrumental in identifying suitable locations for renewable energy projects, such as geothermal or hydroelectric power plants, by assessing the geological conditions of the site. By harnessing natural energy sources, we can reduce our dependence on nonrenewable resources and minimize the environmental impact associated with their extraction and usage.

Furthermore, geological engineering contributes to the preservation and remediation of the environment. The knowledge and expertise of geological engineers are utilized in managing contaminated sites, such as brownfields or mining areas, to minimize the spread of pollutants and restore the affected ecosystems. By developing innovative remediation techniques, geological engineers can mitigate the long-term impact of human activities on the environment, ensuring a sustainable future for generations to come.

In conclusion, the importance of geological engineering cannot be overstated. This field not only ensures the safety and stability of infrastructure projects but also contributes to environmental conservation and sustainable resource management. As students interested in geological engineering, we have the opportunity to make a significant impact on society by utilizing our knowledge to address the challenges we face today. By pursuing a career in this field, we can actively contribute to the betterment of our communities and the world at large.

Career Opportunities in Geological Engineering

Geological engineering is an exciting and dynamic field that offers a wide range of career opportunities for students interested in the intersection of geology and engineering. This subchapter explores the various career paths available to those pursuing a degree in geological engineering, providing valuable insights and guidance for students considering this field of study.

One of the primary career paths for geological engineers is in the mining industry. With their expertise in geology and engineering principles, geological engineers play a crucial role in mineral exploration, extraction, and processing. They are responsible for assessing the geological conditions of potential mining sites, designing safe and efficient mining operations, and ensuring environmental sustainability.

Another area of employment for geological engineers is in the field of environmental consulting. As environmental concerns continue to grow, there is an increasing demand for professionals who can assess and mitigate the impact of human activities on the Earth's geology and environment. Geological engineers contribute their knowledge to projects such as site remediation, groundwater management, and natural hazard assessment.

Geotechnical engineering is yet another field where geological engineers find numerous career opportunities. Geotechnical engineers are involved in the design and construction of infrastructure projects such as bridges, dams, and tunnels. Geological engineers bring their understanding of the subsurface conditions and geologic hazards to ensure the stability and safety of these structures.

Beyond these traditional career paths, geological engineers can also explore opportunities in academia, research institutions, and government agencies. Teaching and research positions allow geological engineers to contribute to the field's knowledge and mentor future generations of students. Government agencies often hire geological engineers for roles related to land-use planning, disaster management, and policy development.

In conclusion, a career in geological engineering opens up a world of exciting opportunities for students passionate about geology and engineering. Whether it's working in the mining industry, environmental consulting, geotechnical engineering, academia, or government agencies, geological engineers have the chance to make a significant impact on society and the environment. By combining their knowledge of geology and engineering principles, they can contribute to sustainable resource extraction, environmental conservation, and the development of safe and resilient infrastructure.

Chapter 2: Fundamentals of Geology

The Earth's Structure and Composition

As students diving into the exciting world of geological engineering, it is crucial to understand the fundamental aspects of our planet's structure and composition. By unraveling the secrets hidden beneath the Earth's surface, we can gain invaluable insights into the dynamics of geological processes and their impact on engineering projects.

The Earth can be divided into several distinct layers based on their composition and physical properties. The outermost layer, known as the crust, is the thinnest layer and consists of solid rock. Beneath the crust lies the mantle, which extends approximately 2,900 kilometers towards the Earth's core. The mantle is primarily composed of solid rock, but it also contains partially molten materials. Finally, at the center of the Earth, we have the core, consisting of two parts: the outer core, which is liquid, and the inner core, which is solid.

Understanding the composition of these layers is crucial for geological engineers. The Earth's crust is comprised of various rocks, including igneous, sedimentary, and metamorphic rocks. These rocks possess different characteristics, such as strength, porosity, and permeability, which directly impact engineering projects such as tunneling, mining, and building foundations.

The mantle, although mostly inaccessible, plays a significant role in geological engineering. Its convective movement drives plate tectonics, resulting in earthquakes, volcanic eruptions, and the formation of mountain ranges. By studying the behavior of the mantle, geological engineers can better understand and predict these geologic hazards, mitigating potential risks to infrastructure and human life.

The Earth's core, despite being beyond our reach, influences the planet's magnetic field, which is crucial for navigation and communication systems. Geological engineers utilize this knowledge to design and implement structures, such as power lines and pipelines, while considering the Earth's magnetic field.

Moreover, the Earth's structure and composition affect the distribution of natural resources, including minerals, fossil fuels, and groundwater. Geological engineers play a vital role in exploring and extracting these resources sustainably, balancing economic development with environmental preservation.

In conclusion, comprehending the Earth's structure and composition is of utmost importance for students of geological engineering. By delving into the secrets hidden within our planet, we gain a deeper understanding of geological processes, enabling us to design and implement engineering projects that are both safe and sustainable. Whether it is building foundations, mitigating geologic hazards, or extracting natural resources, the Earth's structure and composition form the foundation of our work as geological engineers.

Types of Rocks and Minerals

In the fascinating world of geological engineering, it is crucial to understand the various types of rocks and minerals that make up the Earth's crust. These rocks and minerals play a fundamental role in shaping the landscape, the natural resources, and even the geological hazards that engineers must consider in their work. In this subchapter, we will explore the different types of rocks and minerals commonly encountered in geological engineering.

One of the most common types of rocks is sedimentary rock. Formed by the accumulation and compaction of sediments over time, sedimentary rocks hold valuable clues about the Earth's history. Some examples of sedimentary rocks include sandstone, limestone, and shale. Sandstone, as the name suggests, is composed of sand-sized particles that have been cemented together. Limestone, on the other hand, is primarily made up of calcium carbonate and can be found in ancient coral reefs or as the foundation of famous landmarks like the Great Pyramids of Egypt. Shale is a fine-grained sedimentary rock that often contains fossils and can serve as a source of valuable natural gas.

Another category of rocks is igneous rocks. These rocks are formed through the cooling and solidification of molten rock, known as magma or lava. Igneous rocks can be further classified into two types: intrusive and extrusive. Intrusive rocks, such as granite, form beneath the Earth's surface and cool slowly, allowing large mineral crystals to develop. Extrusive rocks, on the other hand, cool rapidly at or near the Earth's surface, resulting in smaller mineral crystals. Examples of extrusive igneous rocks include basalt and obsidian.

Lastly, we have metamorphic rocks. Metamorphic rocks are formed when existing rocks undergo profound changes due to extreme heat and pressure. As a result, the original minerals in the rock recrystallize and form new minerals with different properties. Examples of metamorphic rocks include marble, slate, and gneiss. Marble, known for its beauty and durability, is transformed from limestone through the process of metamorphism. Slate, a fine-grained rock, is commonly used for roofing tiles due to its low water absorption properties. Gneiss, with its banded appearance, is often found in mountain ranges and represents a high degree of metamorphism.

Understanding the characteristics and properties of these rocks and minerals is essential for geological engineers. By identifying and analyzing the different types of rocks and minerals present in a given area, engineers can make informed decisions about construction methods, resource extraction, and even environmental management. So, let's delve deeper into the world of rocks and minerals to unravel the secrets of geological engineering together!

Geological Time Scale

Understanding the concept of time in geological engineering is crucial for students in this field as it provides a framework to comprehend the Earth's history and the processes that have shaped it over millions of years. The geological time scale is a tool that helps us organize and contextualize this vast expanse of time into manageable units.

The geological time scale is divided into several hierarchical units, ranging from the largest eon to the smallest epoch. Let's take a closer look at these divisions.

The largest division is the eon, which represents the longest periods in Earth's history. The current eon, the Phanerozoic, is further divided into three eras: the Paleozoic, Mesozoic, and Cenozoic. Each era is characterized by distinct geological events, such as the emergence of complex life forms during the Paleozoic or the age of dinosaurs in the Mesozoic.

Within each era, we have periods, which are further divided into epochs. These shorter time intervals allow us to study more specific geological events and changes. For example, the Cretaceous period within the Mesozoic era is known for its diverse dinosaur species and the eventual mass extinction event that wiped them out.

The geological time scale is not just a chronological record but also a record of the Earth's changing climate, ecosystems, and the evolution of life. By studying the fossils found in different layers of sedimentary rocks, scientists can identify and date various geological events, such as the appearance of new species or the occurrence of mass extinctions.

Geological engineers rely on the geological time scale to understand the formation and behavior of rocks and minerals. By analyzing the age of specific rock formations, engineers can determine their stability and durability, essential information when planning construction projects or assessing the potential for natural disasters like earthquakes or landslides.

In conclusion, the geological time scale is a fundamental tool for students of geological engineering. It allows us to understand the Earth's history in a systematic and organized manner, providing insights into the processes that have shaped our planet over millions of years. By studying the geological time scale, students can gain a deeper appreciation for the dynamic nature of the Earth and apply this knowledge in their future careers as geological engineers.

Chapter 3: Engineering Geology

Understanding Geological Processes

Geological processes are at the core of the field of geological engineering. As students of this fascinating discipline, it is crucial for us to have a deep understanding of these processes and their implications. This subchapter will delve into the fundamental aspects of geological processes, providing a comprehensive overview that will lay the groundwork for our future studies and practical applications in geological engineering.

To comprehend geological processes, we must first recognize the dynamic nature of our planet. Earth is constantly evolving, shaped by a wide range of natural forces and phenomena. These processes can be broadly categorized into two types: endogenic and exogenic. Endogenic processes arise from within the Earth's interior, such as volcanic activity and tectonic movements, while exogenic processes are driven by external forces like erosion, weathering, and deposition.

One of the primary endogenic processes is plate tectonics, which refers to the movement and interaction of large sections of the Earth's crust. This process gives rise to earthquakes, volcanic eruptions, and the formation of mountain ranges. Understanding plate tectonics is crucial for geological engineers as it directly impacts the stability of structures and the likelihood of natural disasters in specific regions.

Exogenic processes, on the other hand, shape the Earth's surface over long periods of time. Weathering, for example, breaks down rocks into smaller particles through physical and chemical processes. Erosion then carries these particles away, while deposition results in the accumulation of sediments in new locations. These processes have significant implications for geological engineers, particularly in terms of soil erosion, slope stability, and the design of effective drainage systems.

Additionally, this subchapter will explore the impact of geological processes on the availability and quality of natural resources. Geological engineers play a vital role in the extraction and management of Earth's resources, including minerals, fossil fuels, and groundwater. By understanding the geological processes involved in the formation and distribution of these resources, students will develop the necessary skills to contribute to sustainable resource management practices.

In conclusion, understanding geological processes is crucial for students pursuing a career in geological engineering. Through a comprehensive exploration of endogenic and exogenic processes, as well as their implications for the availability and quality of natural resources, this subchapter will provide a solid foundation for our future studies and practical applications in the field. By grasping the intricacies of geological processes, we will be equipped to tackle the complex challenges faced by geological engineers and contribute to the sustainable development of our planet.

Geotechnical Investigation Methods

In the fascinating world of geological engineering, understanding the ground beneath our feet is essential to ensure the safety and stability of structures and infrastructure projects. This subchapter, "Geotechnical Investigation Methods," delves into the various techniques used by geological engineers to investigate and analyze the ground conditions.

Geological engineers employ a range of methods to gather crucial information about the subsurface. One of the primary techniques is known as drilling, where boreholes are created to extract samples of the soil and rock layers. These samples are then meticulously analyzed to determine their composition, strength, and permeability. By studying these properties, geological engineers can accurately assess the behavior and stability of the ground for construction projects.

Another vital method extensively used in geotechnical investigations is geophysical testing. This technique involves the use of seismic waves, electrical resistivity, and ground-penetrating radar to obtain information about the subsurface without physically drilling. Geophysical testing provides valuable insights into the soil and rock layers' properties, helping engineers identify potential hazards such as sinkholes or geological faults.

Furthermore, laboratory testing plays a crucial role in geotechnical investigations. Soil samples obtained from drilling are subjected to a battery of tests to determine their physical and mechanical properties, such as density, moisture content, shear strength, and compressibility. These laboratory tests enable geological engineers to assess the soil's behavior under different loads and environmental conditions, aiding in the design of foundations and earthworks.

In addition to these techniques, geotechnical investigations also involve the monitoring of ground movements. Sophisticated instruments and sensors are used to measure parameters like settlement, slope stability, and groundwater levels. This real-time data allows engineers to detect any changes or anomalies in the ground conditions, helping them make informed decisions during construction projects.

Geotechnical investigations are invaluable in ensuring the safety and success of engineering projects. By employing a combination of drilling, geophysical testing, laboratory analysis, and monitoring, geological engineers gain a comprehensive understanding of the ground's characteristics. This knowledge allows them to design and construct structures that can withstand various geological challenges, including earthquakes, landslides, and soil erosion.

As aspiring geological engineers, it is essential to familiarize ourselves with these investigation methods. By mastering these techniques, we can contribute to the development of sustainable and resilient infrastructure that harmonizes with the dynamic nature of our planet.

Site Characterization and Geological Mapping

In the field of geological engineering, understanding the characteristics of a site and mapping its geological features are paramount for successful project planning and execution. Site characterization and geological mapping are fundamental steps in uncovering the secrets hidden beneath the Earth's surface. For students pursuing a career in geological engineering, these skills are essential for their professional development.

Site characterization involves gathering information about the physical, chemical, and geological properties of a particular location. This process provides engineers with a comprehensive understanding of the site's unique characteristics, enabling them to design and implement appropriate engineering solutions. Geological mapping, on the other hand, involves the identification and documentation of geological formations, structures, and other features within a given area. Together, site characterization and geological mapping provide crucial insights into the subsurface conditions and geologic hazards that may affect engineering projects.

To conduct site characterization, geological engineers employ various techniques such as surface mapping, geophysical surveys, and subsurface exploration. Surface mapping involves the visual inspection and mapping of exposed rock outcrops, soil profiles, and other surface features. Geophysical surveys use specialized instruments to measure and analyze physical properties of the subsurface, like seismic waves or electrical resistivity, to infer the underlying geology. Subsurface exploration techniques, such as drilling and sampling, allow engineers to directly access and analyze soil and rock samples from different depths.

Geological mapping, on the other hand, involves the systematic collection and analysis of geological data to create detailed maps. These maps provide invaluable information about the distribution, nature, and properties of various geological units present in a specific area. By understanding the geological history and structure of the site, engineers can identify potential risks and challenges, such as landslides, earthquakes, or groundwater issues, that may impact their projects. This knowledge allows them to develop effective engineering strategies to mitigate these risks.

For students of geological engineering, learning about site characterization and geological mapping is a stepping stone towards becoming effective professionals in the field. Through theoretical studies, practical exercises, and fieldwork, students gain hands-on experience in mapping geological features and characterizing sites. By understanding the geological context, they can make informed decisions and devise appropriate engineering solutions that are both safe and sustainable.

In conclusion, site characterization and geological mapping are crucial aspects of geological engineering. These practices enable engineers to understand the unique properties and hazards of a site, providing a foundation for successful project planning and execution. For aspiring geological engineers, mastering these skills is essential for their professional development and ensuring the safety and success of future engineering projects.

Chapter 4: Soil Mechanics

Properties of Soils

Understanding the properties of soils is a fundamental aspect of geological engineering. Soils, which are the unconsolidated materials that cover the Earth's surface, play a vital role in various engineering projects. From building foundations to road construction, the behavior of soils directly affects the stability and performance of structures. In this subchapter, we will delve into the key properties of soils that every geological engineering student should be familiar with.

One of the fundamental properties of soils is their composition. Soils are composed of different types of particles, including sand, silt, clay, and organic matter. The proportion of these particles determines the soil's texture, which influences its engineering behavior. For example, sandy soils have larger particles and tend to drain water quickly, while clayey soils have smaller particles and retain water, making them more susceptible to swelling and shrinkage.

Another crucial property of soils is their density. Soil density affects its load-bearing capacity and stability. Compaction is a common method used in engineering to increase soil density. By compacting the soil, engineers can improve its strength and reduce settlement, making it suitable for supporting heavy structures.

In addition to composition and density, soil moisture content is a vital property to consider. The amount of water present in the soil affects its strength, compressibility, and permeability. Excessive moisture can weaken the soil, leading to instability, whereas insufficient moisture can result in soil shrinkage and reduced load-bearing capacity.

Balancing soil moisture is essential in engineering projects to ensure the stability and longevity of structures.

Furthermore, soil shear strength is a critical property for geological engineers to analyze. Shear strength determines the soil's resistance to deformation and failure under applied forces. By conducting shear strength tests, engineers can assess the stability of slopes, embankments, and retaining walls, enabling them to design structures that can withstand potential failure.

Lastly, soil permeability is a property that measures how easily water can flow through the soil. Understanding soil permeability is crucial in managing groundwater and preventing water-related issues, such as soil erosion or the buildup of pore pressure. By conducting permeability tests, geological engineers can assess the suitability of soils for various applications, such as landfill liners or drainage systems.

In conclusion, the properties of soils are essential knowledge for students studying geological engineering. Understanding soil composition, density, moisture content, shear strength, and permeability allows engineers to make informed decisions when designing and constructing structures. By considering these properties, geological engineers can ensure the stability, durability, and safety of engineering projects in various geological environments.

Soil Classification Systems

Understanding the properties and behavior of soil is crucial for geological engineers, as it forms the foundation for various engineering projects like building structures, roads, and dams. Soil classification systems provide a framework for categorizing and organizing different types of soil based on their characteristics, allowing engineers to make informed decisions and design appropriate structures.

One widely used soil classification system is the Unified Soil Classification System (USCS). Developed by the United States Department of Agriculture and the U.S. Army Corps of Engineers, this system classifies soils into several categories based on their physical properties. The USCS considers factors such as grain size distribution, consistency, plasticity, and organic content to classify soils into groups like gravels, sands, silts, and clays. Each group is further divided into subgroups, allowing engineers to identify the specific properties of a particular soil.

Another commonly used classification system is the AASHTO soil classification system, which is specifically designed for highway engineering. This system takes into account the load-bearing capacity and compressibility of soils, classifying them into groups like A-1, A-2, A-3, A-4, A-5, and A-6. Each group represents a different soil type, with A-1 being the most stable and A-6 being the least stable.

Soil classification systems are vital tools for geological engineers as they help in predicting soil behavior, determining the suitability of a site for construction, and selecting appropriate construction techniques. By understanding the properties of different soil types, engineers can make informed decisions regarding foundation design, slope stability, and soil improvement techniques.

It is important for students studying geological engineering to familiarize themselves with these classification systems as early as possible. By understanding the fundamentals of soil classification, students can develop a strong foundation for their future engineering careers. As they progress in their studies, they will learn to analyze soil samples, interpret classification data, and apply this knowledge to real-world engineering problems.

To further enhance their understanding, students can also engage in practical exercises such as soil sampling and laboratory testing. By gaining hands-on experience, they will become proficient in identifying and classifying different soil types, and will be better equipped to tackle the challenges they may encounter in their professional careers.

In conclusion, soil classification systems are essential tools for geological engineers, enabling them to categorize and understand the various properties of soils. By familiarizing themselves with these systems, students of geological engineering can develop a solid foundation for their future careers and make informed decisions when it comes to designing and constructing engineering projects.

Soil Compaction and Consolidation

As students diving into the fascinating world of geological engineering, it is crucial to understand the concepts of soil compaction and consolidation. In this subchapter, we will explore the significance of these processes and their impact on geological engineering projects.

Soil compaction refers to the process of increasing the density of soil by reducing the volume of air within its void spaces. This technique is widely employed in various construction projects, such as building foundations, roadways, and embankments. By compacting soil, engineers can enhance its load-bearing capacity, reduce settlement, and minimize potential structural failures.

The compaction process involves the use of heavy machinery, such as compactors and rollers, to apply pressure on the soil. This pressure expels air from the void spaces, causing the soil particles to come closer together. As a result, the soil becomes denser and more stable, capable of withstanding heavy loads and avoiding undesirable settlement.

Consolidation, on the other hand, refers to the gradual process of soil settling over time due to the expulsion of water from its void spaces. It occurs when a load is applied to a saturated soil, causing an increase in the effective stress and subsequent water drainage. Consolidation is a vital consideration in geological engineering, particularly in projects involving soft or compressible soils, such as clay deposits.

The consolidation process is influenced by various factors, including the soil's permeability, compressibility, and the applied load. Understanding these factors is essential for predicting the settlement of structures and designing appropriate foundations. Engineers employ methods like consolidation tests and laboratory experiments to determine the soil's compressibility and estimate the time it takes for consolidation to occur.

By comprehending the concepts of soil compaction and consolidation, geological engineering students can effectively analyze the behavior of different soil types and design structures that can withstand the forces acting upon them. The knowledge gained from these processes enables engineers to make informed decisions in selecting appropriate construction techniques and materials.

In conclusion, soil compaction and consolidation play vital roles in the field of geological engineering. Understanding these processes allows students to comprehend the behavior of soils under various conditions and design structures that can withstand the forces imposed upon them. By implementing proper soil compaction techniques and considering consolidation, engineers can ensure the stability and longevity of their projects. So, delve into the secrets of soil compaction and consolidation and unlock the key to successful geological engineering endeavors.

Chapter 5: Rock Mechanics

Behavior of Rocks under Stress

Understanding the behavior of rocks under stress is a fundamental aspect of geological engineering. As students pursuing a career in geological engineering, it is essential to comprehend the response of rocks to external forces and the implications this has on various engineering projects. This subchapter will delve into the fascinating world of rock behavior under stress, exploring the different types of stress that rocks experience, how they deform, and the associated engineering implications.

Rocks, being solid materials, possess a certain degree of strength and rigidity. However, when subjected to stress, they exhibit different behaviors depending on the magnitude and duration of the applied forces. There are three primary types of stress that rocks experience: compression, tension, and shear. Compression occurs when rocks are squeezed together, tension when they are stretched apart, and shear when they slide past each other.

When rocks are subjected to compressive stress, they tend to deform by either folding or faulting. This deformation can significantly impact the stability of engineering structures such as tunnels, dams, and buildings. Understanding the behavior of rocks under compression is crucial for designing and constructing these structures to withstand the anticipated loads and avoid catastrophic failures.

Tensional stress, on the other hand, leads to the development of fractures or cracks in rocks. These fractures can propagate and weaken the overall integrity of the material. Geological engineers must consider the potential for tension-induced fractures when designing underground mining operations, rock slopes, and foundations to ensure their stability and safety.

Shear stress results in the sliding or movement of rocks along planes known as faults. This shear movement can cause rock masses to shift, leading to landslides and slope failures. Geological engineers employ various techniques to analyze shear behavior, such as shear strength tests, to identify potential failure zones and implement appropriate stabilization measures.

Understanding the behavior of rocks under stress also involves examining their response to cyclic loading, temperature variations, and water saturation. The interaction between these factors and rock behavior is crucial for assessing the long-term stability of geological structures and designing effective mitigation strategies.

In conclusion, the behavior of rocks under stress plays a central role in geological engineering. By comprehending how rocks deform and respond to different types of stress, students can effectively design and execute engineering projects while ensuring the safety and sustainability of structures. This knowledge is invaluable for the successful practice of geological engineering and the advancement of the field as a whole.

Rock Strength and Failure Criteria

Understanding the strength and failure criteria of rocks is fundamental to the field of geological engineering. As students in this exciting discipline, it is essential for us to grasp the concepts and theories behind rock strength and failure criteria as they form the basis for designing safe and efficient engineering projects in the realm of geological engineering.

Rock strength refers to the ability of a rock material to withstand external forces without undergoing permanent deformation or failure. It is a crucial parameter in assessing the stability of rock masses and designing structures such as tunnels, dams, and foundations. The strength of rocks is influenced by various factors, including mineral composition, grain size, porosity, and geological history.

In geological engineering, we use failure criteria to determine the conditions under which rocks will fail and deform permanently. Failure criteria are mathematical models that help us understand how rocks behave under different stress conditions. By analyzing the stress-strain relationships, we can predict the failure modes and potential failure mechanisms of rock masses.

There are several common rock failure criteria used in geological engineering. The most widely used criterion is the Mohr-Coulomb failure criterion, which considers both the shear strength and normal stress acting on a rock mass. Another important criterion is the Hoek-Brown criterion, which takes into account the rock's strength properties and geological characteristics, providing a more accurate prediction of rock behavior.

To determine the strength and failure criteria of rocks, various laboratory tests are conducted. These tests include uniaxial compression tests, triaxial compression tests, and direct shear tests, among others. Through these tests, we can measure the rock's strength parameters, such as compressive strength, tensile strength, and shear strength, which are crucial for designing engineering structures.

As students of geological engineering, we must also consider the geological conditions and environmental factors that affect rock strength and failure criteria. These factors include geological formations, weathering processes, and groundwater conditions. By integrating geological knowledge with engineering principles, we can make informed decisions and develop effective strategies to mitigate potential risks and ensure the stability and safety of our engineering projects.

In conclusion, rock strength and failure criteria are central to the field of geological engineering. By understanding and applying these concepts, we can assess the stability of rock masses and design structures that can withstand the forces exerted upon them. As students of this discipline, it is vital for us to grasp these theories, conduct laboratory tests, and consider the geological context to become competent and successful geological engineers.

Slope Stability Analysis

When it comes to geological engineering, one of the most critical aspects to consider is slope stability. Understanding and analyzing the stability of slopes is crucial for ensuring the safety and functionality of various engineering projects, such as highways, dams, and buildings. In this subchapter, we will delve into the fascinating world of slope stability analysis, providing students with the essential knowledge and tools to tackle this complex subject.

To begin with, let's define slope stability. A slope is considered stable when it can resist the forces that act upon it, such as gravity, water pressure, and external loads. However, various factors can compromise slope stability, leading to potentially disastrous consequences. Therefore, it is vital for geological engineers to thoroughly assess the stability of slopes before and during construction.

In this subchapter, we will introduce students to the different methods used for slope stability analysis. We will explore both the analytical and numerical techniques employed in the field, highlighting their strengths and limitations. Students will gain an understanding of the principles behind each method, allowing them to choose the most suitable approach based on the specific project requirements.

Furthermore, we will discuss the key parameters and factors that influence slope stability. These include soil properties, groundwater conditions, slope geometry, and external loads. Students will learn how to collect and analyze data related to these factors, enabling them to make accurate assessments of slope stability.

Moreover, we will delve into the various failure mechanisms that can occur in slopes, such as shallow and deep-seated landslides, slope creep, and slope deformation. By understanding these failure mechanisms, students will be able to identify potential risks and develop effective mitigation measures.

Lastly, we will explore the role of geotechnical instrumentation in slope stability analysis. Students will learn about the different monitoring techniques used to assess slope behavior and detect early warning signs of instability. This knowledge will equip them with the skills required to implement effective monitoring programs and ensure the long-term stability of slopes.

In conclusion, slope stability analysis is a fundamental aspect of geological engineering. By studying and understanding the principles, methods, and factors involved in slope stability analysis, students will be well-equipped to tackle challenges in their future careers. This subchapter aims to provide a comprehensive overview of slope stability analysis, empowering students to make informed decisions and contribute to the safe and sustainable development of engineering projects.

Chapter 6: Geotechnical Design

Foundation Engineering

Foundation Engineering is a crucial aspect of Geological Engineering that focuses on the design and construction of solid and stable foundations for various structures. This subchapter aims to provide students with a comprehensive understanding of the principles and techniques involved in this field.

In the world of Geological Engineering, foundation plays a vital role in the stability and safety of any structure. Whether it is a skyscraper, a bridge, or a simple residential building, a strong and well-designed foundation is essential for the longevity and integrity of the structure. Foundation Engineering involves the analysis of soil and rock properties to determine the most suitable foundation type for a particular site.

One of the key topics covered in this subchapter is site investigation. Students will learn about the importance of conducting thorough investigations to assess the geological conditions of a site. This includes studying the soil composition, groundwater levels, and any potential hazards such as landslides or seismic activities. By understanding the site's properties, students can make informed decisions about the foundation design.

The subchapter also delves into different types of foundations, such as shallow and deep foundations. Students will learn about the design principles for each type, including factors like load-bearing capacity, settlement, and lateral stability. Various techniques, such as soil improvement and ground reinforcement, will be explored to enhance the foundation's performance.

Furthermore, the subchapter covers the analysis and design of retaining structures, which are used to support and stabilize slopes. Students will gain insights into the different types of retaining walls and their applications, along with the design principles and factors to consider in their construction.

Throughout the subchapter, case studies and real-world examples will be provided to illustrate the practical applications of foundation engineering principles. Students will also be introduced to the latest technologies and software used in the field, allowing them to develop essential technical skills.

By the end of this subchapter, students will have a solid understanding of foundation engineering and its significance in Geological Engineering. They will be equipped with the knowledge and skills necessary to analyze soil properties, design appropriate foundations, and ensure the stability and safety of structures in various geological conditions.

Retaining Structures and Earthworks

In the fascinating world of geological engineering, retaining structures and earthworks play a crucial role in shaping the landscape and ensuring the stability of various construction projects. As students entering the realm of geological engineering, it is essential to grasp the fundamental concepts and techniques associated with these vital components.

Retaining structures refer to man-made structures designed to hold back soil, rock, or other materials and prevent slope failures. These structures are commonly employed in projects such as highways, bridges, and buildings that are situated on sloping terrain. Understanding the principles behind retaining structures is of utmost importance, as they are instrumental in maintaining the integrity and safety of these structures.

One of the key factors to consider when designing retaining structures is the type of soil or rock being retained. Different materials have distinct characteristics, such as cohesion, angle of repose, and drainage properties, which must be taken into account during the design process. By studying the properties of various soil and rock types, geological engineering students can determine the most suitable retaining structure design for a given project.

Earthworks, on the other hand, encompass a wide range of activities that involve the manipulation of soil and rock for construction purposes. This can include excavations, embankments, and grading. Understanding earthworks is vital for geological engineering students, as it allows them to effectively alter the topography of a site to suit construction requirements.

When undertaking earthworks, it is vital to consider factors such as slope stability, compaction, and erosion control. By carefully analyzing the soil and rock properties, students can determine the most appropriate techniques for excavation or embankment construction. Moreover, implementing erosion control measures helps mitigate the potential environmental impacts of earthworks, ensuring sustainable engineering practices.

To enhance your knowledge in this field, it is essential to familiarize yourself with various retaining structure designs, including gravity walls, cantilever walls, and sheet pile walls. Additionally, understanding the principles behind earthworks, such as the importance of proper compaction and slope stability analysis, will equip you with the necessary tools to tackle real-world projects.

As students venturing into the captivating world of geological engineering, mastering the concepts and techniques associated with retaining structures and earthworks is paramount. By delving into the intricacies of these topics, you will be well-prepared to tackle the challenges that lie ahead in your geological engineering career.

Ground Improvement Techniques

In the field of geological engineering, ground improvement techniques play a crucial role in addressing various challenges related to the stability and load-bearing capacity of soil and rock formations. These techniques aim to enhance the properties of the ground, making it suitable for construction projects and ensuring the safety and longevity of structures.

One commonly used ground improvement technique is compaction. This process involves the densification of loose or poorly compacted soils by applying mechanical energy. Compaction reduces the air voids within the soil, increasing its density and improving its load-bearing capacity. This technique is often employed in the construction of roads, embankments, and foundations for buildings.

Another effective ground improvement technique is soil stabilization. This method is used to strengthen weak or unstable soils by adding stabilizing agents such as cement, lime, or fly ash. These agents react with the soil particles, improving its engineering properties and preventing settlement or slope instability. Soil stabilization is particularly useful in areas with soft clay or loose sand, where the natural soil lacks the necessary strength for construction.

Geosynthetics, such as geotextiles and geogrids, are also valuable tools in ground improvement. These synthetic materials are used to reinforce and stabilize soil, preventing erosion and enhancing its load-bearing capacity. Geotextiles are often used in road construction to separate different soil layers and prevent mixing, while geogrids are employed in retaining walls and slopes to provide additional strength and stability.

In certain situations, ground improvement techniques may involve ground freezing. This method is typically used to stabilize soil or rock formations during excavation or tunneling projects. By injecting refrigerants into the ground, the soil or rock becomes frozen, creating a temporary support structure. This technique is particularly useful in areas with high groundwater levels or unstable soil conditions.

It is important for students in the field of geological engineering to familiarize themselves with these ground improvement techniques. By understanding and applying these methods, students can contribute to the successful and safe completion of construction projects. Moreover, gaining expertise in ground improvement techniques enables engineers to address the unique geological challenges faced in different regions, contributing to sustainable and resilient infrastructure development.

Chapter 7: Environmental Geotechnics

Contaminant Transport in Soil and Groundwater

In the field of geological engineering, understanding the transport of contaminants in soil and groundwater is of utmost importance. As students pursuing a career in geological engineering, it is crucial to comprehend the mechanisms behind the movement of pollutants and their potential impact on the environment. This subchapter aims to shed light on the fascinating world of contaminant transport in soil and groundwater, providing you with essential knowledge and insights.

Contaminants can enter the soil and groundwater through various sources, such as industrial activities, improper waste disposal, or accidental spills. Once introduced into the environment, these pollutants can travel through the soil and eventually reach the groundwater, potentially posing a significant threat to water quality and human health. Therefore, it becomes imperative for geological engineers to assess and mitigate the risks associated with contaminant transport.

The transport of contaminants in soil and groundwater is influenced by several factors, including soil properties, hydrogeological conditions, and the characteristics of the contaminant itself. Understanding these factors and their interplay is fundamental in determining the fate and transport of pollutants.

This subchapter will explore various mechanisms of contaminant transport, including advection, dispersion, diffusion, and adsorption. Advection refers to the movement of contaminants with the flowing groundwater, while dispersion involves the spreading of contaminants due to the heterogeneity of the subsurface. Diffusion plays a role in the movement of contaminants from areas of high concentration to low concentration. Additionally, adsorption occurs when contaminants attach to soil particles, reducing their mobility.

Moreover, we will delve into the techniques and models used by geological engineers to predict and simulate contaminant transport. These tools enable us to assess the potential risks associated with contamination and develop effective remediation strategies. We will discuss the use of numerical models, such as the finite element method, to simulate contaminant transport in complex geological settings.

By understanding contaminant transport in soil and groundwater, you will be equipped with the knowledge and skills to contribute to the protection and preservation of our environment. As future geological engineers, it is our responsibility to ensure the sustainable management of soil and groundwater resources, safeguarding the well-being of both present and future generations.

In conclusion, this subchapter provides an in-depth exploration of contaminant transport in soil and groundwater, offering valuable insights into the world of geological engineering. By studying this topic, you will gain a comprehensive understanding of the mechanisms, factors, and tools involved in assessing and mitigating the risks associated with environmental contamination. Through your knowledge and expertise, you can make a positive impact on the field of geological engineering and contribute to the sustainable development of our planet.

Landfill Design and Management

In the field of geological engineering, one crucial aspect that requires careful consideration is landfill design and management. Landfills are engineered facilities designed to safely dispose of solid waste, minimizing its impact on the environment and human health. As students pursuing a degree in geological engineering, it is essential to understand the principles and practices involved in landfill design and management.

The design of a landfill involves various factors, including site selection, waste characterization, liner system design, leachate collection and treatment, gas management, and final cover design. Site selection is a critical step that involves assessing the geotechnical and hydrogeological properties of the area to ensure the landfill's stability and prevent contamination of groundwater. Waste characterization is necessary to identify the nature and composition of the waste materials, helping engineers determine the appropriate design parameters.

The liner system plays a crucial role in preventing the migration of contaminants into the surrounding soil and groundwater. It typically consists of a composite barrier, including a low-permeability geomembrane, compacted clay liner, and geosynthetic clay liner. These components provide a barrier to the downward movement of leachate, which is the liquid that results from the decomposition of waste materials.

Leachate collection and treatment systems are essential for managing the liquid waste generated within the landfill. These systems include pipes and collection sumps that capture the leachate and transport it to a treatment facility. Proper leachate management is vital to prevent contamination of groundwater and nearby surface water bodies.

Another significant aspect of landfill design and management is gas management. As waste decomposes, it produces landfill gas, mainly comprised of methane and carbon dioxide. These gases can be collected and utilized as a source of renewable energy or safely vented to the atmosphere. However, if not managed properly, landfill gas can pose a significant risk, leading to explosions, fires, and greenhouse gas emissions.

Lastly, the final cover design ensures the long-term stability and environmental protection of the landfill after it reaches its capacity. The final cover typically includes a combination of soil, vegetation, and engineered components to prevent water infiltration, control erosion, and promote long-term stability.

As students of geological engineering, it is crucial to develop a comprehensive understanding of landfill design and management principles. By incorporating sound engineering practices, we can contribute to the responsible and sustainable management of solid waste, protecting both the environment and human health.

Remediation of Contaminated Sites

As students of geological engineering, it is vital for us to understand the importance of remediating contaminated sites. Environmental pollution caused by human activities poses significant threats to the health of ecosystems and the well-being of communities. The remediation of contaminated sites is a critical process that aims to restore the affected areas to their original state or at least minimize the adverse effects of pollution. In this subchapter, we will explore the various techniques and strategies used in the remediation of contaminated sites.

One of the primary methods employed in site remediation is soil and groundwater remediation. Soil contamination can occur due to the release of hazardous substances, such as industrial waste or chemical spills. Groundwater, on the other hand, can become contaminated through the infiltration of pollutants into the soil, threatening drinking water supplies. To address these issues, techniques like excavation and removal, soil vapor extraction, and bioremediation are commonly used. These methods involve the physical removal of contaminated soil, the extraction of volatile compounds, and the use of microorganisms to break down pollutants, respectively.

In addition to soil and groundwater remediation, the remediation of contaminated sites also involves the treatment of air pollution. Air pollution can result from various sources, including industrial emissions and the burning of fossil fuels. To combat this, technologies such as air filtration systems, scrubbers, and catalytic converters are implemented to remove harmful pollutants from the air. These strategies help to improve air quality and reduce the negative impacts on human health and the environment.

Another important aspect of site remediation is the assessment and monitoring of contaminated sites. Before implementing any remediation measures, thorough site investigations are conducted to determine the extent and nature of contamination. This includes sampling and analysis of soil, water, and air samples. Furthermore, continuous monitoring is necessary to evaluate the effectiveness of remediation efforts and ensure that the site is restored to an environmentally safe condition.

In conclusion, the remediation of contaminated sites is an essential task for geological engineers. By understanding and implementing various techniques and strategies, we can contribute to the restoration and protection of the environment. As future professionals in the field, it is crucial for us to stay updated with the latest advancements and research in remediation technologies. By doing so, we can actively participate in the sustainable management of contaminated sites, ensuring a healthier and cleaner future for all.

Chapter 8: Geological Hazards and Risk Assessment

Natural Hazards: Earthquakes, Landslides, and Volcanoes

As students of geological engineering, it is crucial to have a comprehensive understanding of the natural hazards that we may encounter in our field of work. Earthquakes, landslides, and volcanoes are some of the most significant natural events that shape the Earth's surface, and they require our attention and expertise.

Earthquakes, often referred to as the ground shaking phenomenon, occur when there is a sudden release of energy in the Earth's crust. Their occurrence is usually associated with tectonic plate boundaries, where immense forces build up over time. As geological engineers, we play a crucial role in assessing the potential risks of earthquakes and designing structures that can withstand their destructive power. By studying the seismic behavior of different materials and evaluating the geotechnical aspects of construction sites, we can contribute to creating safer and more resilient infrastructures.

Landslides are another significant natural hazard that geological engineers must consider. They occur when masses of rock, soil, or debris move down a slope, often triggered by heavy rainfall, earthquakes, or human activities. Understanding the geological and geotechnical factors that contribute to landslides is essential for mitigating their impact. By analyzing the stability of slopes, designing appropriate drainage systems, and implementing effective slope reinforcement techniques, we can minimize the risk of landslides and protect communities and infrastructure.

Volcanoes, while fascinating natural phenomena, can also be hazardous. These geological features release molten rock, ash, and gases, posing risks to both human life and the environment. As geological engineers, we have the expertise to assess volcanic hazards, monitor volcano activity, and design protective measures to mitigate the potential damage caused by volcanic eruptions. By understanding the behavior of volcanic materials and their interactions with the surrounding environment, we can contribute to the development of effective strategies for volcanic hazard management.

In conclusion, as students of geological engineering, it is crucial for us to understand the natural hazards of earthquakes, landslides, and volcanoes. By studying these phenomena, we can develop the knowledge and skills necessary to assess, mitigate, and manage the risks associated with these hazards. Our expertise in geological engineering allows us to contribute to the creation of safer and more resilient communities, ensuring that infrastructure is designed to withstand the forces of nature.

Risk Assessment and Mitigation Strategies

Risk assessment and mitigation strategies are essential components of the field of geological engineering. As students in this exciting field, it is crucial to understand the importance of identifying potential risks and developing effective strategies to mitigate them. This subchapter will delve into the key concepts of risk assessment and mitigation strategies specific to the field of geological engineering.

Risk assessment involves evaluating and analyzing potential hazards and their associated impacts on engineering projects. In geological engineering, these hazards can range from geological formations, such as landslides and earthquakes, to environmental factors like climate change and water availability. By conducting thorough risk assessments, geological engineers can identify potential risks and estimate their likelihood and potential consequences.

Once risks are identified, mitigation strategies can be developed to minimize or eliminate their impact. These strategies include both preventive measures and reactive approaches. Preventive measures involve implementing design modifications, such as reinforcing structures to withstand potential hazards, or conducting thorough geological surveys to identify potential risks before undertaking a project. Reactive approaches, on the other hand, focus on developing contingency plans and emergency response strategies to effectively manage unforeseen risks.

One common risk mitigation strategy in geological engineering is the use of monitoring systems. These systems track and analyze geological conditions and provide real-time data to engineers, allowing them to detect any potential risks and take immediate action. For example, monitoring systems can detect ground movement, enabling engineers to predict and prevent landslides or other geological hazards.

Geological engineers also rely on geotechnical investigations to assess the stability of soil and rock formations. By conducting detailed surveys and laboratory tests, engineers can determine the strength and behavior of these materials, enabling them to design structures that can withstand potential risks.

In addition to technical strategies, effective risk communication is crucial in geological engineering. Students need to understand the importance of clear and concise communication with stakeholders, such as clients, government agencies, and local communities. By effectively communicating risks and mitigation strategies, geological engineers can ensure that all parties involved understand the potential challenges and are prepared to address them.

In conclusion, risk assessment and mitigation strategies are fundamental in the field of geological engineering. By conducting thorough risk assessments, implementing preventive measures, and developing reactive approaches, geological engineers can minimize the impact of potential hazards on engineering projects. Through the use of monitoring systems, geotechnical investigations, and effective risk communication, students in geological engineering can play a crucial role in ensuring the safety and success of engineering projects in the face of geological risks.

Case Studies of Geological Hazards

In the field of geological engineering, it is crucial to understand the various geological hazards that can occur and their potential impact on the environment and human life. By studying real-life case studies, students of geological engineering can gain valuable insights into the causes, effects, and mitigation strategies related to different geological hazards. This subchapter presents a collection of compelling case studies that highlight the significance of geological hazards and their implications for geological engineering.

One case study focuses on the devastating earthquake that struck the city of Kobe, Japan, in 1995. This earthquake, measuring 6.9 on the Richter scale, caused extensive damage to infrastructure and claimed thousands of lives. Students will explore the geological factors that contributed to the severity of this earthquake, such as the location of fault lines and the type of soil in the area. They will also learn about the innovative engineering techniques implemented during the city's reconstruction to enhance its resilience against future earthquakes.

Another case study delves into the catastrophic landslide that occurred in Oso, Washington, in 2014. This massive landslide, triggered by heavy rainfall, engulfed an entire neighborhood, resulting in the loss of numerous lives and extensive property damage. By examining this case study, students will gain a deep understanding of the geological conditions that made the area prone to landslides and the challenges faced by geological engineers in mitigating the risks associated with such hazards.

The subchapter also includes a case study on the eruption of Mount St. Helens in Washington state in 1980. This volcanic eruption, classified as a VEI-5 event, caused widespread destruction and resulted in the loss of several lives. Students will explore the volcanic processes leading up to the eruption, including the buildup of pressure and the release of ash and pyroclastic flows. They will also examine the measures taken by geological engineers to monitor volcanic activity and protect nearby communities from future eruptions.

By studying these and other case studies, students of geological engineering will gain a comprehensive understanding of the complexities and challenges associated with geological hazards. They will learn how geological engineers play a crucial role in assessing risks, designing safe infrastructure, and developing effective mitigation strategies to protect communities and the environment. These case studies will not only enhance students' technical knowledge but also inspire them to contribute to the field by developing innovative solutions to mitigate geological hazards.

Chapter 9: Geotechnical Laboratory and Field Testing

Common Laboratory Tests for Geotechnical Engineering

In the field of geological engineering, laboratory testing plays a crucial role in understanding the properties of soil and rock formations. These tests enable engineers to assess the stability and load-bearing capacity of structures, as well as evaluate the potential for soil erosion or landslides. As students pursuing a degree in geological engineering, it is essential to familiarize yourselves with the common laboratory tests used in this field. This subchapter aims to provide you with a comprehensive overview of these tests.

One of the fundamental laboratory tests in geotechnical engineering is the Atterberg limits test. This test determines the moisture content at which a soil transitions between different states, namely liquid, plastic, and solid. By understanding the Atterberg limits, engineers can assess the soil's consistency and determine its suitability for construction projects.

Another critical test is the grain size analysis, which determines the distribution of particle sizes in a soil sample. This test helps engineers classify soils into various categories, such as sand, silt, or clay. The grain size analysis is essential in designing foundations, selecting construction materials, and predicting soil behavior.

In geotechnical engineering, the consolidation test is used to measure the compression characteristics of soil. This test examines how a soil sample consolidates under increasing pressure and provides valuable information about settlement and stability. By understanding consolidation, engineers can design foundations that can withstand the anticipated settlement of the soil.

Shear strength is a vital property of soil that determines its resistance to deformation under load. The direct shear test is commonly employed to measure shear strength. This test involves applying a controlled shear force to a soil sample and observing its response. Understanding a soil's shear strength is crucial for designing slopes, retaining walls, and other earth structures.

The triaxial test is another laboratory test that assesses the shear strength and stress-strain behavior of soil. It applies axial and radial stresses to a soil sample under controlled conditions, allowing engineers to analyze its behavior in complex stress states. The triaxial test is particularly important in the design of deep foundations and underground structures.

These are just a few examples of the common laboratory tests used in geotechnical engineering. By familiarizing yourselves with these tests, you will gain a solid foundation in understanding soil behavior and its implications for engineering projects. Mastering these laboratory techniques will enable you to make informed decisions and contribute significantly to the field of geological engineering.

In-situ Testing Techniques

Geological engineering is a fascinating field that encompasses the exploration and utilization of Earth's resources for various engineering projects. As students delving into this subject, it is essential to understand the significance of in-situ testing techniques. These methods play a pivotal role in evaluating the properties and behavior of soil and rock formations, providing valuable insights for geological engineering projects.

In-situ testing involves conducting experiments and measurements directly within the natural environment, eliminating the need for laboratory simulations. This approach allows engineers to obtain accurate data, reflecting the actual conditions of a site. By employing in-situ testing techniques, students can gain a comprehensive understanding of the geological characteristics and geotechnical properties of the Earth's materials, enabling them to make informed decisions during the planning and execution of engineering projects.

One commonly used in-situ testing technique is the cone penetration test (CPT). This method involves pushing a cone-shaped probe into the ground and measuring the resistance encountered. The data obtained from the CPT can provide valuable information about soil composition, density, and strength. This information is crucial for designing foundations, assessing slope stability, and determining the bearing capacity of soils.

Another important in-situ testing technique is the pressuremeter test. By inserting an expandable device into the ground and gradually increasing the pressure, engineers can measure the deformation characteristics of the soil or rock. This test allows students to assess the compressibility and strength of the materials, helping them design safe and stable structures.

Furthermore, seismic testing techniques, such as the seismic refraction and seismic reflection methods, are widely used in geological engineering. These methods involve generating and measuring seismic waves to determine the subsurface geological composition and stratigraphy. By analyzing the velocity and reflection patterns of the waves, students can identify potential risks, such as fault lines or unstable ground conditions, and devise appropriate engineering solutions.

In conclusion, in-situ testing techniques are indispensable tools for geological engineering students. These methods enable them to gather accurate data about the properties and behavior of soils and rocks, providing valuable insights for various engineering projects. By mastering these techniques, students can make informed decisions, design safe structures, and contribute to the sustainable development of our built environment.

Data Interpretation and Analysis

In the field of Geological Engineering, data interpretation and analysis play a crucial role in understanding and predicting geological phenomena. As students of this fascinating discipline, it is essential for us to develop proficiency in these skills to excel in our studies and future careers.

Data interpretation involves extracting meaningful insights from raw data collected through various geological surveys and experiments. This step requires a deep understanding of geological principles and concepts, as well as the ability to identify relevant patterns and trends. By interpreting data, we can gain valuable information about the Earth's composition, structure, and processes. This knowledge is essential for making informed decisions in the field of geological engineering, such as assessing the stability of slopes, designing foundations for structures, or mitigating the impact of natural disasters.

Once data is interpreted, the next step is data analysis. This involves applying statistical methods and techniques to further analyze and validate the interpreted information. Statistical analysis allows us to quantify uncertainties, identify outliers, and establish correlations between different variables. By employing these analytical tools, we can evaluate the reliability and accuracy of our data interpretations, ensuring that our conclusions are based on sound scientific principles.

To effectively interpret and analyze geological data, it is crucial to utilize appropriate software and programming tools. Familiarity with programs such as MATLAB, Python, and GIS (Geographical Information System) can greatly enhance our ability to handle large datasets and perform complex calculations. These tools enable us to visualize data in meaningful ways, create detailed maps, and develop models that simulate real-world geological scenarios.

Furthermore, effective communication of data interpretation and analysis is vital in the field of geological engineering. As students, we must be able to present our findings and results in a clear and concise manner. This involves using visual aids such as graphs, charts, and diagrams to effectively communicate complex geological concepts to our peers and industry professionals.

In conclusion, data interpretation and analysis are essential skills for students pursuing a career in geological engineering. By developing proficiency in these areas, we can enhance our understanding of geological phenomena and make valuable contributions to the field. Through the use of appropriate software and effective communication, we can ensure that our interpretations and analyses are accurate, reliable, and accessible to a wider audience.

Chapter 10: Introduction to Geotechnical Software

Overview of Geotechnical Software Applications

Geotechnical software applications play a vital role in the field of geological engineering. These powerful tools have revolutionized the way we analyze, design, and manage various geotechnical projects. In this subchapter, we will provide an overview of the different geotechnical software applications commonly used in the field, highlighting their significance and benefits.

One of the most widely used software applications in geological engineering is geotechnical modeling software. These programs allow engineers to create accurate and detailed three-dimensional models of subsurface conditions. By inputting geotechnical data such as soil properties, groundwater levels, and geological formations, engineers can simulate real-world scenarios and analyze the behavior of the soil and rock masses. This helps in making informed decisions regarding site investigations, foundation design, and slope stability analysis.

Another important software application is geotechnical analysis software. These programs utilize various numerical methods to solve complex geotechnical problems. Engineers can perform stability analyses for slopes, retaining walls, and deep excavations, considering factors such as soil properties, groundwater conditions, and loading conditions. These software applications also aid in designing foundation systems, assessing bearing capacities, and predicting settlements. The ability to quickly analyze and evaluate different design options significantly improves the efficiency and accuracy of geological engineering projects.

In addition to modeling and analysis, geotechnical software applications also assist in project management and documentation. Documentation software helps in organizing and storing geotechnical data, including borehole logs, laboratory test results, and site photographs. This ensures easy access to crucial information, facilitating collaboration among team members and enhancing project efficiency. Furthermore, project management software allows engineers to plan and track project timelines, allocate resources, and monitor project progress. These tools help streamline project workflows and ensure successful project completion.

For students pursuing a career in geological engineering, it is crucial to familiarize themselves with these geotechnical software applications. These tools are extensively used in the industry and provide a competitive advantage in the job market. By gaining proficiency in geotechnical software applications during their academic journey, students can develop strong analytical and problem-solving skills, enabling them to tackle real-world geotechnical challenges effectively.

In conclusion, geotechnical software applications are invaluable tools for geological engineering. From geotechnical modeling and analysis to project management and documentation, these software applications enhance the efficiency, accuracy, and overall success of geotechnical projects. As students, it is essential to embrace these tools and develop proficiency in their usage to excel in the field of geological engineering.

Hands-on Training with Geotechnical Software

In the field of geological engineering, hands-on experience with geotechnical software is crucial for developing a strong foundation in the discipline. As students, it is important to familiarize ourselves with the various tools and techniques that will be essential in our future careers. This subchapter aims to provide a comprehensive overview of the significance of hands-on training with geotechnical software and its role in shaping our understanding of geological engineering.

Geotechnical software plays a pivotal role in the analysis and design of geotechnical structures, such as foundations, retaining walls, and slopes. By engaging in hands-on training with these software programs, students gain practical knowledge of the tools used by professionals in the field. This practical experience allows us to enhance our problem-solving skills and develop a deep understanding of the complexities involved in geological engineering projects.

One of the key advantages of hands-on training with geotechnical software is the ability to simulate real-world scenarios. Through these simulations, students can evaluate the behavior of geotechnical structures under different conditions and analyze the potential risks and challenges they may face. By experimenting with various design parameters and inputting different soil properties, we can gain vital insights into the performance and stability of these structures.

Furthermore, hands-on training enables students to grasp the importance of data interpretation and analysis. Geotechnical software provides us with vast amounts of data, ranging from soil properties to groundwater levels. By learning how to interpret and analyze this data, we can make informed decisions and recommendations for engineering projects. This skill is crucial in effectively communicating our findings and collaborating with other professionals in the field.

Hands-on training with geotechnical software also fosters a sense of confidence and competence among students. As we become proficient in using these tools, we feel empowered to tackle complex engineering problems and propose innovative solutions. This practical experience instills a sense of professionalism and prepares us for the challenges we may encounter in our future careers.

In conclusion, hands-on training with geotechnical software is a vital component of a student's journey in geological engineering. By engaging in this training, we acquire practical skills, simulate real-world scenarios, and develop a deep understanding of the complexities involved in geotechnical projects. It empowers us to make informed decisions, analyze data effectively, and confidently tackle engineering challenges. Ultimately, this hands-on experience equips us with the necessary tools to excel in the field of geological engineering.

Importance of Computer-Aided Design in Geological Engineering

In today's rapidly evolving technological landscape, computer-aided design (CAD) has become an indispensable tool for professionals in various fields, including geological engineering. This subchapter aims to shed light on the significance of CAD in the context of geological engineering, providing students with a comprehensive understanding of its applications and benefits within their chosen niche.

Geological engineering is a specialized field that deals with the study of the Earth's subsurface, aiming to mitigate risks associated with natural hazards, design sustainable infrastructure, and explore valuable resources. The complexity and scale of geological projects make CAD an invaluable asset for geological engineers, facilitating efficient planning, accurate analysis, and effective communication of design concepts.

One of the primary advantages of CAD in geological engineering is its ability to create detailed and precise 3D models of geological formations. Through the use of specialized software, students can bring complex geological data to life, visualizing subsurface structures, rock types, and other critical parameters. These models allow for a deeper understanding of the geological context, enabling engineers to make informed decisions and develop optimized designs.

Furthermore, CAD software enables students to simulate and analyze various geological processes, such as groundwater flow, slope stability, and seismic activity. By inputting relevant data into the software, students can accurately predict potential hazards, assess risks, and devise appropriate mitigation strategies. This not only enhances the safety and stability of engineering projects but also minimizes potential environmental impacts.

CAD also plays a crucial role in fostering collaboration among multidisciplinary teams. Geological engineering projects often require collaboration with professionals from diverse fields, such as civil engineering, geology, and environmental science. CAD tools provide a common platform for these experts to exchange information, share design iterations, and address any conflicts or discrepancies. This seamless integration of knowledge and resources enhances the overall quality and efficiency of geological engineering projects.

In conclusion, computer-aided design has revolutionized the field of geological engineering, empowering students to tackle complex challenges with greater precision and efficiency. From creating detailed 3D models to simulating geological processes and fostering collaboration, CAD has undoubtedly become an indispensable tool for aspiring geological engineers. By embracing CAD technology, students can unlock new possibilities, improve project outcomes, and contribute to the sustainable development of our planet.

Chapter 11: Case Studies in Geological Engineering

Geotechnical Challenges in Construction Projects

Construction projects come with their fair share of challenges, especially when it comes to the discipline of geological engineering. In this subchapter, we will explore some of the geotechnical challenges that students pursuing a career in geological engineering may encounter in construction projects.

One of the primary challenges in construction projects is the assessment and understanding of the site's geology. Geological engineers must study the composition, structure, and properties of the earth's materials to ensure the stability and safety of the construction site. This involves conducting extensive surveys and tests to identify potential hazards such as unstable soil, rock formations, or groundwater presence.

Another significant challenge is the management of soil and rock excavations. Geological engineers play a crucial role in determining the appropriate excavation methods and ensuring the stability of the surrounding area during the process. They must consider factors such as slope stability, soil erosion, and the potential for landslides or collapses.

Furthermore, construction projects often involve the design and construction of foundations, which can be particularly challenging in geologically complex areas. Geological engineers must analyze the soil and rock properties to determine the most suitable foundation type, such as shallow foundations, deep foundations, or specialized techniques like ground improvement. They also need to consider factors such as settlement, bearing capacity, and potential differential movements that can affect the stability of the structure.

In addition to the geological aspects, construction projects may also face challenges related to environmental and sustainability concerns. Geological engineers must consider the impact of construction activities on the surrounding ecosystems, natural resources, and the local community. They need to develop strategies to minimize environmental damage, ensure proper waste management, and promote sustainable construction practices.

To overcome these challenges, students pursuing geological engineering must acquire a strong foundation in geotechnical principles, geological mapping, and site investigation techniques. They should also develop proficiency in using advanced software and modeling tools to analyze and predict the behavior of geological materials.

In conclusion, geotechnical challenges are an integral part of construction projects, and geological engineers play a vital role in addressing these challenges. By understanding the site's geology, managing excavations, designing appropriate foundations, and considering environmental factors, geological engineers contribute to the safe and sustainable development of construction projects. As students in the field of geological engineering, it is essential to develop a comprehensive understanding of these challenges to become successful professionals in the industry.

Rehabilitation of Damaged Structures

Title: Rehabilitation of Damaged Structures: Restoring Stability in Geological Engineering

Introduction:

The field of geological engineering plays a crucial role in ensuring the stability and resilience of structures built on or near geological formations. However, natural disasters and unforeseen events can cause significant damage to these structures, posing challenges that require immediate attention. In this subchapter, we will explore the rehabilitation techniques used to restore damaged structures, with a focus on their application in the field of geological engineering. Whether you are a student or professional in the field, understanding these techniques is essential for safeguarding the built environment.

1. Assessing Structural Damage:
Before initiating any rehabilitation efforts, it is crucial to conduct a comprehensive assessment of the damaged structure. This involves identifying the type and extent of damage, understanding the underlying geological conditions, and evaluating potential risks.

2. Strengthening Techniques:
Structural strengthening techniques play a pivotal role in rehabilitating damaged structures. The subchapter will delve into various techniques, such as external bonding, steel plate bonding, and fiber-reinforced polymers (FRP) – their advantages, limitations, and suitability for different geological conditions.

3. Soil Stabilization:
In many geological engineering projects, soil stability is a critical concern. Damaged structures often require soil stabilization to prevent

further settlement or movement. This section will explore different methods, including grouting, soil nailing, and ground improvement techniques like deep soil mixing and dynamic compaction.

4. Foundation Rehabilitation: Foundations are the backbone of any structure, and their rehabilitation is of utmost importance. The subchapter will discuss techniques such as underpinning, micropiles, and jet grouting to restore the integrity of damaged foundations.

5. Seismic Retrofitting: Given the seismic nature of geological engineering, structures must be retrofitted to withstand earthquakes and other seismic events. This section will cover techniques like base isolation, energy dissipation devices, and strengthening of structural elements to enhance seismic performance.

6. Case Studies: To provide practical insights, this subchapter will feature real-world case studies showcasing successful rehabilitation projects in geological engineering. These case studies will demonstrate the application of various techniques and their effectiveness in restoring damaged structures.

Conclusion:
Rehabilitating damaged structures in geological engineering is a multifaceted process that requires a deep understanding of structural behavior, geological conditions, and available rehabilitation techniques. By comprehending the principles and applications discussed in this subchapter, students and professionals in geological engineering can contribute to the preservation and resilience of our built environment.

Innovations in Geological Engineering

Geological engineering is a fascinating field that combines the principles of geology and engineering to address various challenges related to the Earth's surface and subsurface. As students of this discipline, it is crucial to stay updated with the latest innovations that are shaping the future of geological engineering. In this subchapter, we will explore some of the most exciting advancements in this field, giving you a glimpse into the cutting-edge technologies and techniques that are revolutionizing the industry.

One of the most significant innovations in geological engineering is the use of remote sensing technologies. These tools, such as LiDAR (Light Detection and Ranging) and satellite imagery, allow engineers to collect data from large areas quickly and accurately. This data can then be analyzed to identify potential geological hazards, evaluate the stability of slopes, or determine the suitability of a site for construction projects. By harnessing the power of remote sensing, geological engineers are now able to make informed decisions based on comprehensive and up-to-date information.

Another remarkable innovation is the development of advanced geotechnical modeling software. These computer programs simulate real-world geological conditions, enabling engineers to predict the behavior of soils and rocks under different loads and environmental conditions. Through these models, engineers can optimize the design of foundations, tunnels, and other structures, minimizing risks and ensuring their long-term stability. Such software also facilitates the analysis of complex geological phenomena, such as landslides and earthquakes, aiding in the development of effective mitigation strategies.

In recent years, the field of geological engineering has also seen advancements in the area of sustainable and environmentally friendly practices. With the growing concern for the planet's well-being, engineers are now working on innovative techniques to minimize the environmental impact of their projects. These include the use of geothermal energy for heating and cooling systems, the implementation of green infrastructure to manage stormwater runoff, and the integration of ecological considerations in land development planning. By embracing these sustainable practices, geological engineers are playing a vital role in creating a more resilient and eco-friendly future.

In conclusion, the field of geological engineering is constantly evolving, driven by innovative technologies and the need for sustainable solutions. As students of this discipline, it is crucial to embrace these advancements and stay up-to-date with the latest developments. By doing so, we can contribute to the field's growth and be prepared to tackle the complex challenges that lie ahead. So, let's continue unearthing the secrets of geological engineering and embark on a journey of innovation and discovery.

Chapter 12: Future Trends in Geological Engineering

Emerging Technologies in the Field

In the ever-evolving field of geological engineering, staying up-to-date with the latest advancements is crucial for students pursuing a career in this niche. As technology continues to transform various industries, geological engineering is no exception. This subchapter aims to shed light on some of the emerging technologies that are revolutionizing the field of geological engineering, providing students with a glimpse into the future of their profession.

1. Remote Sensing and Geographical Information Systems (GIS): Remote sensing techniques, such as satellite imagery and LiDAR, are proving to be valuable tools for geological engineers. These technologies allow for the collection of vast amounts of geospatial data, providing a comprehensive understanding of geological structures, terrain, and potential hazards. Coupled with GIS, this data can be analyzed, visualized, and interpreted to make informed decisions regarding engineering projects and land management.

2. Unmanned Aerial Vehicles (UAVs) and Drones: UAVs and drones have gained popularity in recent years, and their application in geological engineering is no exception. Equipped with high-resolution cameras and sensors, these devices can capture aerial imagery and collect data in remote or hazardous areas. By utilizing UAVs, geological engineers can conduct detailed surveys, monitor environmental changes, and inspect infrastructure without compromising safety.

3. 3D Printing and Rapid Prototyping: 3D printing has revolutionized the manufacturing industry, and its

impact is increasingly felt in geological engineering. This technology allows for the creation of intricate and customized models, prototypes, and even tools. Geological engineers can now better visualize their designs, assess structural integrity, and optimize their projects before implementation. 3D printing also enables the development of lightweight and cost-effective materials, reducing environmental impact and improving sustainability.

4. Artificial Intelligence (AI) and Machine Learning: The vast amounts of data generated by geological engineering projects can be efficiently processed and analyzed using AI and machine learning algorithms. These technologies can identify patterns, detect anomalies, and predict geological events. By leveraging AI, geological engineers can enhance decision-making, optimize resource allocation, and improve the accuracy of geological hazard assessments.

5. Robotics and Autonomous Systems: Robotic technologies are increasingly being employed in geological engineering for tasks such as mapping, exploration, and monitoring. Autonomous systems equipped with sensors and cameras can navigate challenging terrains and collect data in real-time. These technologies not only enhance productivity but also mitigate risks associated with human intervention in hazardous environments.

As students, embracing these emerging technologies is paramount to staying ahead in the field of geological engineering. Familiarizing oneself with these advancements will not only broaden career prospects but also equip students with the necessary tools to tackle future challenges in a rapidly changing world. By embracing innovation and integrating these technologies into their skill set, students can become the future leaders and problem solvers in the field of geological engineering.

Sustainable Practices in Geological Engineering

In recent years, the field of geological engineering has been increasingly focused on sustainability and environmental stewardship. As students in this discipline, it is vital for us to understand and implement sustainable practices in our work to ensure the long-term health of our planet.

One of the key principles of sustainable geological engineering is the responsible management of natural resources. As geological engineers, we often work with minerals, water, and energy resources. It is our duty to develop and implement techniques that minimize waste and ensure the efficient use of these resources. This involves employing innovative technologies and strategies to reduce the environmental impact of our operations, such as adopting efficient drilling and extraction methods, and implementing water and energy conservation practices.

Another crucial aspect of sustainable geological engineering is the consideration of environmental impacts. We must strive to minimize the disruption to ecosystems and habitats during our projects. This can be achieved by conducting thorough environmental impact assessments before commencing any work, and implementing mitigation measures to protect wildlife and preserve biodiversity. Additionally, we can explore alternative and less invasive methods, such as geophysical surveys, to minimize the need for extensive excavation and reduce the disturbance to natural landscapes.

Sustainable practices in geological engineering also encompass the responsible management of waste and pollution. It is our responsibility to develop and implement strategies for the safe disposal and treatment of hazardous materials generated during our operations. Recycling and reusing materials whenever possible can also help reduce waste and minimize the demand for new resources.

Furthermore, as students of geological engineering, we should advocate for sustainable practices within the industry. This involves staying informed about the latest research and technological advancements in sustainability, and actively engaging in discussions and collaborations with professionals and peers. By promoting sustainable practices and raising awareness about their importance, we can contribute to the growth of a more environmentally conscious field.

In conclusion, sustainable practices in geological engineering are crucial for the well-being of our planet and the future of our profession. As students, it is our responsibility to fully understand and implement these practices in our work. By adopting resource-efficient techniques, minimizing environmental impacts, managing waste and pollution responsibly, and advocating for sustainability, we can contribute to a more sustainable and resilient geological engineering industry. Let us embrace sustainability and become agents of positive change in our field.

Opportunities for Research and Development

In the field of geological engineering, the opportunities for research and development are vast and exciting. As students pursuing a career in this niche, you will have the chance to explore and unearth the secrets of our planet, while also contributing to the advancement of society. This subchapter will outline some of the most promising opportunities for research and development in geological engineering, inspiring you to embark on a journey of discovery and innovation.

One of the key areas of research in geological engineering is the exploration and extraction of mineral resources. As the demand for minerals continues to rise, engineers are constantly seeking more efficient and sustainable methods of extraction. Through research, you can contribute to developing new technologies that minimize the environmental impact of mining operations, such as the use of advanced robotics, machine learning, and remote sensing techniques.

Another exciting area of research is geotechnical engineering, which focuses on understanding and mitigating risks associated with natural hazards like earthquakes, landslides, and volcanic eruptions. By studying the behavior of soils and rocks under different conditions, you can contribute to the development of innovative techniques for designing and constructing structures that can withstand these hazards. This research not only ensures the safety of human lives but also protects infrastructure and contributes to the overall resilience of communities.

Climate change is another pressing issue that offers numerous research opportunities in geological engineering. As our planet experiences rapid changes, scientists and engineers are working together to understand the impacts of climate change on geological systems and develop strategies to mitigate its effects. From studying the melting of polar ice caps to assessing the risks of coastal erosion, your research can contribute to finding sustainable solutions to protect our environment and communities.

In addition to these specific areas, geological engineering research also encompasses topics such as groundwater management, environmental remediation, and geological hazard mapping. The interdisciplinary nature of this field allows you to collaborate with experts from various backgrounds, including geologists, geophysicists, environmental scientists, and civil engineers. This collaboration fosters a rich learning environment and promotes innovative thinking.

As students, you have the opportunity to shape the future of geological engineering through your research and development endeavors. By exploring these diverse areas, you will not only expand your knowledge but also contribute to the sustainable development and protection of our planet. Take advantage of these opportunities, embrace the challenges, and let your curiosity lead the way as you uncover the secrets of geological engineering.

Chapter 13: Conclusion and Final Thoughts

Reflecting on the Journey as a Geological Engineering Student

As a student pursuing a degree in geological engineering, your academic journey is an exciting and challenging one. This subchapter aims to provide you with a reflective exploration of what it means to be a geological engineering student, the experiences you may encounter, and the significant impact this field can have on society.

From the moment you embark on this path, you will find yourself immersed in a world where nature and technology merge. Geological engineering offers a unique perspective on the Earth's structure, geologic processes, and the utilization of these resources to address various engineering challenges. As a geological engineering student, you will delve into subjects such as geology, soil mechanics, rock mechanics, groundwater flow, and geotechnical engineering. Each class will contribute to your understanding of the Earth's dynamics and its implications for engineering projects.

Throughout your journey, fieldwork will emerge as an essential aspect of your education. Geological engineering students spend countless hours outdoors, observing and analyzing geological formations, collecting soil and rock samples, and conducting geotechnical investigations. These hands-on experiences will deepen your understanding of the theories learned in the classroom, while also fostering critical thinking and problem-solving skills.

Moreover, the interdisciplinary nature of geological engineering opens doors to a wide range of career opportunities. As you progress in your studies, you will realize that geological engineering is not limited to construction sites or mining operations. It extends into environmental consulting, geothermal energy, natural hazard assessment, and even space exploration. The skills and knowledge gained as a geological engineering student will equip you to tackle the challenges our planet faces, such as climate change, resource management, and sustainable development.

While the journey as a geological engineering student may be demanding, it is also incredibly rewarding. The ability to bridge the gap between the natural world and engineering solutions is both intellectually stimulating and fulfilling. Moreover, the impact of your work can be felt on a global scale, as geological engineers contribute to the development of infrastructure, the mitigation of natural hazards, and the preservation of our environment.

In conclusion, reflecting on your journey as a geological engineering student allows you to appreciate the unique blend of science and engineering this field offers. Embrace the challenges, immerse yourself in fieldwork, and explore the diverse career opportunities available to you. By doing so, you will not only become a skilled professional but also contribute to the advancement of geological engineering and the betterment of our planet.

Advice for Future Geological Engineering Students

Congratulations on choosing to pursue a career in geological engineering! This subchapter aims to provide valuable advice and insights to help you navigate your journey and make the most of your time as a geological engineering student. Whether you are just starting your academic journey or are already immersed in your studies, these tips will help you excel in your field.

1. Embrace the interdisciplinary nature: Geological engineering is a multidisciplinary field that combines geology, engineering, and environmental science. Embrace this interdisciplinary nature and seek opportunities to learn from various disciplines. This will enhance your problem-solving skills and make you a well-rounded professional.

2. Develop a strong foundation in mathematics and science: As a geological engineering student, you will encounter complex mathematical and scientific concepts. It is crucial to build a strong foundation in math and science subjects to excel in your coursework and future career.

3. Engage in hands-on experiences: Take advantage of opportunities to engage in fieldwork, laboratory experiments, and internships. These experiences will provide you with practical skills and a deeper understanding of geological engineering concepts. Additionally, they will help you build a network of professionals in the field.

4. Join professional societies and organizations: Get involved in geological engineering societies and organizations to connect with like-minded individuals and experts in the field. These networks will offer you access to resources, mentorship, and potential job opportunities.

5. Seek guidance from faculty and professionals: Develop relationships with your professors and professionals in the field. They can provide valuable guidance, mentorship, and career advice. Attend office hours, participate in research projects, and seek opportunities to collaborate with your instructors.

6. Stay updated with industry trends: Geological engineering is a rapidly evolving field. Stay updated with the latest industry trends, technological advancements, and research findings. This will not only give you a competitive edge but also help you understand the potential challenges and opportunities that lie ahead.

7. Cultivate strong communication skills: Effective communication is essential as a geological engineering professional. Work on improving your written and oral communication skills. Practice presenting your work and research findings to diverse audiences to enhance your ability to convey complex ideas concisely.

8. Take advantage of resources and libraries: Familiarize yourself with the resources available to you, such as geological databases, research journals, and libraries. These resources will be invaluable throughout your academic and professional journey.

9. Stay curious and never stop learning: Geological engineering is a field that constantly evolves. Stay curious, ask questions, and never stop learning. Seek opportunities for continuous professional development and stay updated with the latest research and industry practices.

Remember, your journey as a geological engineering student is not just about acquiring knowledge but also about developing critical thinking skills and a passion for the field. Embrace the challenges, persevere through the difficult times, and always strive for excellence. Good luck on your path to becoming a successful geological engineering professional!

Continuing Education and Professional Growth in Geological Engineering

In the field of Geological Engineering, the pursuit of knowledge and professional growth is not limited to the years spent in classrooms and laboratories. As students of Geological Engineering, we must recognize the importance of continuing education to stay updated with the latest advancements and to foster our personal and professional growth.

Continuing education in Geological Engineering is a lifelong process that allows us to expand our expertise, gain new skills, and explore emerging areas within our field. It enables us to adapt to the ever-evolving challenges faced by the industry, as well as contribute to innovative solutions that address environmental concerns, natural hazards, and sustainable development.

One of the most effective ways to engage in continuing education is through professional conferences and workshops. These events bring together experts, researchers, and professionals from various backgrounds, providing us with a platform to learn from their experiences and exchange knowledge. Attending these conferences allows us to stay updated with the latest research findings, technological advancements, and industry best practices. Furthermore, it offers opportunities to network with professionals, which can lead to collaborations and career opportunities.

Another avenue for continuing education is through online courses and webinars. With the advent of technology, we now have access to a wealth of resources at our fingertips. These online platforms provide us with the flexibility to learn at our own pace, from anywhere in the world. Topics ranging from geotechnical engineering, environmental geology, to remote sensing and GIS, can be explored through these courses. By investing our time in these resources, we can enhance our knowledge and skills, making us more competitive in the job market.

Furthermore, professional certifications and licenses play a crucial role in the career advancement of a Geological Engineer. Acquiring certifications such as Professional Geologist (P.G.) or Engineer-In-Training (EIT) demonstrates our commitment to professionalism and adherence to ethical standards. These certifications not only enhance our credibility but also open doors to higher positions and increased job opportunities.

In conclusion, continuing education and professional growth are vital aspects of a successful career in Geological Engineering. By actively engaging in lifelong learning, attending conferences, participating in online courses, and obtaining professional certifications, we can stay at the forefront of the industry, contribute to its growth, and achieve our full potential. Embracing these opportunities will not only benefit us as individuals but also contribute to the advancement and sustainability of our field.